TRIUMPH
TOURING TWINS
3T - 5T - 6T - 3TA - 5TA

1938 - 1966

Roy Bacon

NITON PUBLISHING

TRIUMPH TOURING TWINS

First published in the United Kingdom by:
Niton Publishing
PO Box 3 . Ventnor . Isle of Wight PO38 2AS

Acknowledgements
The author would like to thank those who
helped this book by supplying the photographs.
Most came from the EMAP archives or *Motor
Cycle News* by courtesy of Editor Malcolm
Gough. Others came from the Mick Woollett
archive and some from the author's files.

Filmset and printed by Crossprint
Newport . Isle of Wight

ISBN 1 85579 003 3
A CIP catalogue record for this book is available
from the British Library

Front Cover: The 1948 Speed Twin as depicted on the front cover of the brochure
for that year and capturing the spirit of the time so well.

Back Cover: The unit construction model Twenty-One or 3TA, complete with its
bathtub, as shown in the 1961 Triumph brochure.

Ethel Revnell seated on a police Speed Twin at Radiolympia late in 1949.

Contents

Introduction

Triumph launched their Speed Twin model at Earls Court in September 1937, after its announcement in the previous July. It was not that firm's first vertical twin, but it was the one that set the trend for the British industry in the post-war years, and which all the major firms were to copy.

The Speed Twin was followed by the sports Tiger 100 in 1939 and, post-war, these two were joined by the smaller 3T. Later, came the larger Thunderbird which, in turn, was developed into the high-performance Tiger 110 and then the Bonneville. The 3T was not built for long, but the other models were developed further during the 1950s.

In 1957, the first of the small unit-construction twins appeared as the Model 21 and offered considerable enclosure of the rear part of the machine. This set a trend, just as the original Speed Twin had done in 1937, and for a while it seemed that this would be the way forward. Two years later, it was joined by a larger version, which replaced the original Speed Twin, but kept both its name and traditional colour.

These two ran on up to 1966, together with the Thunderbird, which was enclosed at the rear for 1960 and then turned to unit-construction in 1963. However, the trend reversed to sports models and café racers, and to follow this the enclosure was reduced and eventually removed altogether.

The touring Triumph twins had their own place in motorcycling, as they offered good performance for everyday riding. With this came economy and reliability, plus behaviour on the road that was easy to live with from day to day. There was no real power band or cam-effect to make riding difficult when conditions were tricky on the journey home after a hard day. The discerning mourned their passing, and the models proved as popular as ever in the classic revival of the 1980s.

Try-out of a TT marshal's Thunderbird in the Isle of Man in 1956 when the machines were fitted with radio equipment and an alternator as well as the dynamo to provide the power.

Speed Twin

Triumph built their first vertical twin back in 1913, after trials with a Bercley engine in 1909, and another, designed by Val Page, in 1933. The latter engine was produced as the model 6/1 for three years, but bore little relationship to what was to become the archetypal Triumph twin that set the industry standard.

The firm had gone through a difficult period in the mid-1930s, after which Edward Turner became chief designer and general manager. He was autocratic and demanding, but made things happen fast in the company, which was to benefit greatly from his flair for styling and knack of spotting trends. This com-pounded into an uncanny ability to know just how far to go to be the trend-setter, while keeping within the limits of what the public would buy.

Turner soon showed this by revamping the rather staid, but worthy, Val Page singles to produce the Tiger models. These had sparkle, shine and glamour, so they were an immediate success as sports machines, having a combination of style, finish and good basic design.

Amidst all the work needed to establish the firm soundly in its restructured form, Turner found time to design the machine he is best remembered for. It was announced

A Smith Illuminated Chronometric Trip Speedometer (120 m.p.h.) will be supplied unless otherwise ordered, £2-15-0 extra.

The original Speed Twin as shown in the 1938 brochure, the start of a long, long line.

Drive side of the Speed Twin for 1939 when the extra block fixings and front number plate surround appeared.

as the model T, but soon became the Speed Twin and model 5T. As with most of his work, the design was simple, small and light, following the philosophy that real weight reduction had to be carried out at the drawing board.

In fact, the overhead-valve engine was lighter than that of the single-cylinder Tiger 90, of the same 500 cc capacity, and narrow enough to fit into the same frame. It was built much as a single, so had few extra parts, and its lines followed those of a twin-port single, then a popular style despite the extra cost of the second exhaust system.

Turner chose to use a vertical twin, as it fitted so neatly into the existing cycle parts and avoided the carburation and other problems inherent in V- and flat-twin engines. The next step along that line of thought was to select a 360 degree twin, but really this followed logically on most counts. The type gave even firing intervals, which allowed a standard twin-cylinder magneto to be used and a single carburettor on a short manifold. Thanks to the more constant depression in the carburettor, a smaller instrument could be used, and any blow-back tended to be taken by the other cylinder.

The even firing gave a good exhaust note, always important to riders, and the engine torque was smoother than that of a single. Engine balance was slightly better, thanks to the shorter stroke, but was not to be a problem for the touring engines, which did not run fast enough for vibration to be noticed much. Later, with more capacity, power and engine speed, it was to create problems.

So, the Turner twin had vertical cylinders and a 360 degree crankshaft with its overall layout much as for a single. The engine dimensions were 63 x 80 mm, the same as the

Postwar Speed Twin of 1947 with telescopic front forks and the optional sprung hub at the rear.

firm's 250 cc models, and the capacity was 499 cc. The design was based on a crankcase that was split vertically on the engine centre-line and a one-piece cylinder block, plus separate one-piece cylinder head.

The crankcase halves were cast in aluminium and located to each other by a spigot that ran round the major diameter and a single dowel. A ring of fastenings held the two halves together. These included a pair of screws that went in the crankcase mouth to improve its rigidity

The smaller 3T twin as shown in the 1948 brochure; the model having been introduced in 1945.

and support for the block. An opening at the base of the crankcase assembly was machined flat for a sump plate and filter assembly, which was held by four bolts. The timing-side crankcase half was extended to the rear to provide a platform for the mag-dyno, and the timing chest, formed with the casting, also ran back to enclose the drive to the electrics.

There was a ball-race main bearing in each crankcase half, the larger being fitted on the drive side, plus two phosphor-bronze camshaft bushes located high in the castings, fore and aft of the crankshaft centre-line. Those on the timing side were flanged.

The crankshaft was built up from two steel stampings, which were bolted together with the flywheel sandwiched between them. Each stamping comprised a mainshaft, bobweight, crankpin and inner flange, the machining being common except for the ends of the mainshafts. The inner flanges were spigoted into a recess on each side of the flywheel, and six small high-tensile bolts held the assembly together. The crankpins were hollow to enable lubricating oil to pass through them, and the space between the flanges acted as a sludge trap.

Connecting rods with split big-ends were used, and each rod was

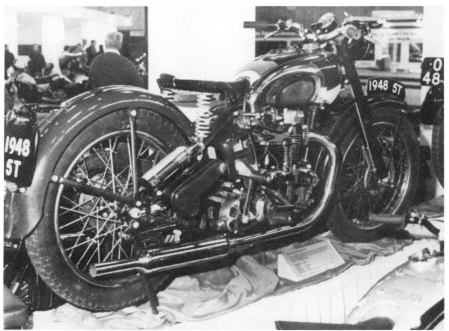

Interesting sectioned Speed Twin of 1948 on show in 1982 when the classic revival was well underway.

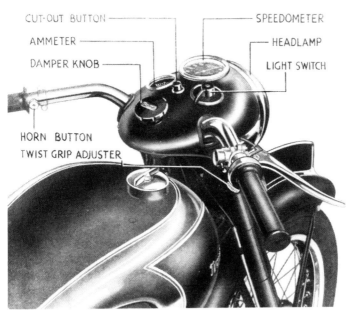

CUT-OUT BUTTON

AMMETER

DAMPER KNOB

SPEEDOMETER

HEADLAMP

LIGHT SWITCH

HORN BUTTON
TWIST GRIP ADJUSTER

The famous
Triumph nacelle
which cleaned up
the front end so
well and again set a
trend for the
industry.

forged in light-alloy and ran directly on the crankpin. Its cap was forged in steel and, in this case, the bearing surface was coated with white metal. The cap was formed with integral studs that ran up through the rod where they wei ɔ secured by castellated nuts, locked with split pins. The small-end eye was fitted with a phosphor-bronze bush.

Both pistons had three rings, including the scraper, and were held by hollow gudgeon pins retained by wire circlips. The compression ratio was 7.0:1, and the pistons had solid skirts and flat tops with small valve cut-aways. They moved in a cast-iron cylinder block, which was held to the crankcase by just six studs and nuts, short cylinder spigots protruding down into the case mouth.

Each cylinder bore was spigoted into the one-piece, cast-iron cylinder head, which was held down to the block by eight bolts, a gasket being fitted between the two. The cylinder head had parallel inlet ports, but the exhaust ports were splayed outwards to assist their cooling, and each terminated in a stub for an exhaust pipe. Wells were cast in the top of the cylinder head for the valves, and a shouldered valve guide was pressed into each. Each valve was controlled by duplex coil springs, which sat on a cup located on the guide, and each was retained by a collar and split collets. A hardened cap went on top of each valve stem.

An aluminium manifold was bolted to the rear of the cylinder head and carried a Type 76 Amal

The nacelle was introduced for 1949 and is seen here on a 3T which was little changed in other respects.

carburettor of 15/16 in. bore. An air filter was listed as an option. On the exhaust side, each port stub had an exhaust pipe fixed to it by a finned clamp, while a tubular silencer was mounted low down on each side of the machine.

There were two separate rocker boxes cast in light-alloy. Each straddled two of the valve wells and contained two rockers on a single long spindle, spring and hardened washers being used to take up end-play and absorb wear. The spindles were hollow and supplied with lubricant from their right-hand ends. The rockers were single forgings, the adjuster screw and its locknut

The four-bar tank style came in for 1950 along with the Thunderbird which brought in improvements shared with the other models.

Period picture taken late in 1949 of Triumph tester Alex Scobie with his 6T talking to Jaguar tester Ron Sutton with the new XK120.

being in the outer arm and a ball-headed pin in the inner arm. Each rocker box had two caps to provide access to the adjuster screws.

The pushrods were positioned close to the centre-line of the engine, to fore and aft of the block, the cooling fins of which were shaped to allow this. Each pushrod ran down to a tappet, which moved in a housing, itself a tight fit in a hole in the base of the block, where it was secured by a single set screw. There were two tappets in each housing and, as they had radiused feet, they worked against each other to pre-

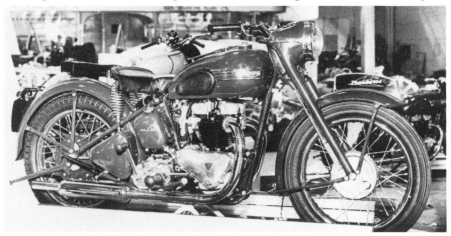

The Thunderbird being introduced at Earls Court late in 1949 when it was part of the 1950 range with the new tank styling.

Sectioned Triumph which was on show at the Festival of Britain Exhibition held at London's South Bank during 1951.

vent their rotation.

A chrome-plated tube ran between each tappet housing and the underside of the rocker box to enclose the pushrods. A pipe assembly was connected to each tube and ran up to the valve wells so that oil could drain from them. For the exhaust wells, the drains were taken from the front of the cylinder head, but for the inlet wells they came from the sides to avoid the manifold. Once in the pushrod tubes, the oil was free to lubricate the tappets in their housing and then the cams themselves.

The camshafts were mounted high in the crankcase, and it has been said that this high camshaft layout, together with the short pushrods and combustion chamber shape, came from the Riley Nine car engine. Certainly, Turner had owned one of these machines, which were highly acclaimed for their advanced design in the 1930s.

The timing gear was of the simplest, a pinion on the crankshaft driving an intermediate gear that ran on a fixed spindle pressed into the crankcase wall. This gear was meshed with gears keyed to the right-hand ends of the camshafts, that of the inlet being meshed with a further gear fitted to the mag-dyno.

The gears were a tight fit on the

Early 1950s police Speed Twin fitted with flashing amber lights used when escorting other vehicles.

camshafts and were held in place by nuts. The nut on the inlet camshaft was formed with an offset pin that drove the oil pump. This was bolted to the crankcase wall within the timing chest and had twin plungers for the dry-sump system with its external oil tank. Pipes connected the tank and engine, which had a suction tube within the crankcase to lift the returning oil from the sump to the pump.

The timing gears and oil pump were enclosed by a cover that gave the engine a line that was unique to Triumph and which it was never to lose totally. On the cover went a small, triangular badge, held by three hammer-drive screws, which identified the model. The cover included a bush, in which ran the extreme right-hand end of the crankshaft, with the pressure release valve of the oil system outboard of it. This controlled the oil flow into the crankshaft end and, thus, to the big-end bearings, while a pressure line from the front of the timing cover supplied the rocker spindles and was connected to a pressure gauge set in the petrol tank instrument panel.

Ignition was by a Lucas type MN2 magneto with an E3HM dynamo strapped to its back. The ignition advance was manually operated by a control on the handlebars, so the drive was by a simple pinion. The dynamo output was regulated by a constant-voltage control box, which went under the seat.

to the left-hand crankshaft end. From the sprocket, a single-strand chain took the power to the clutch, which had four plates with cork inserts, five plain plates and four clutch springs.The clutch was typically Triumph and common with other models. It had a centre for the plates that was splined to a hub mounted on a taper on the gearbox mainshaft. This hub acted as the inner race for the rollers on which

Last year for the 3T was 1951, after which the firm concentrated on the larger twins for some years.

The power output of the engine was quoted as 27 bhp by Edward Turner in 1939, which put it 3 to 4 bhp above the Tiger 90 single, while the engine weight of the twin was marginally lower. Aside from any weight or power advantage, the twin accelerated better thanks to its lighter engine components, and the result was a more responsive machine.

Primary transmission began with a cam-type shock absorber splined

the clutch drum, with integral sprocket, ran. The clutch was lifted by a simple lever mounted in the gearbox outer cover.

The whole of the primary drive was enclosed by a cast-aluminium chaincase, the two halves of which were held together by a row of small screws. In its original form, this had a round boss to clear the crankshaft-mounted shock absorber but, in production,this was faired back in a much nicer style that was to

First year for the SU carburettor on the Thunderbird was 1952 when the frame was amended to clear the air filter hose.

remain for many years. The outer case half had a screwed cap near its top to enable oil to be poured in and the chain tension to be checked.

A stock Triumph four-speed gearbox was used with standard ratios, although it was feasible to convert this to the wide-ratio set. The gear-box was typically British, with the clutch on the mainshaft, outboard of the final drive sprocket on the sleeve gear, and the layshaft beneath the mainshaft. There were ball races to support the sleeve gear and right-hand end of the mainshaft, with bushes used elsewhere.

Six Los Angeles graduates over in England for a European touring holiday with Thunderbird machines.

The gears were selected by forks that slid on a rod under the control of a camplate. This plate was turned by a quadrant, which was moved by the positive-stop mechanism. The foot gear pedal moved in the up-for-up sense, usual for Triumph, and went on the right-hand side of the machine. The kickstart pedal was on the same side, and its shaft carried a quadrant that engaged with a gear on the right-hand end of the mainshaft. This gear drove the shaft via a spring-loaded face ratchet, and both pedals had rubbers fitted to them. The gearbox shell was an aluminium casting, open to the right, and sealed by inner and outer covers, the latter with an oil filler cap.

Nearly everything else of the machine was common to the larger single-cylinder models. The engine and gearbox were mounted in plates that were special to the twin, but the frame was stock. Conventional, brazed lug-and-tube construction was used for the frame, which was built up in two sections. The front comprised the top, seat and down tubes plus headstock, while the rear consisted of the pairs of upper and lower chainstays, the latter extending forward to the bottom of the downtube.

The front forks were stock girders with a single, central spring and adjustable friction dampers on the lower front spindle. A steering damper was fitted, the friction discs being beneath the lower fork yoke and the control knob above the upper. Rubber bushes helped to

The 1953 Speed Twin which had coil ignition and an alternator, but in this case has retained the timing cover for the dynamo drive.

The ignition coil of the 1953 5T went immediately above the distributor and made checking the points that much harder.

isolate the handlebars and their mountings from any vibration.

Both wheels had offset hubs with 7 in. single-leading-shoe brakes, and both front and rear hubs turned on taper-roller races. A quickly-detachable rear wheel was available as an option and used similar bearings. The tyre sizes were given as 26 x 3.0 in.ribbed front, and 26 x 3.5 in. Universal rear, which equated to the 3.00 x 20 in. and 3.50 x 19 in. of later years. The rear brake was connected directly by rod to its control pedal on the left-hand side of the machine.

Adequate mudguards were fitted as standard, with well valanced ones an option. The rear stay of the front mudguard doubled as the front stand, while the rear mudguard was in two sections so that part of it could be detached to assist wheel

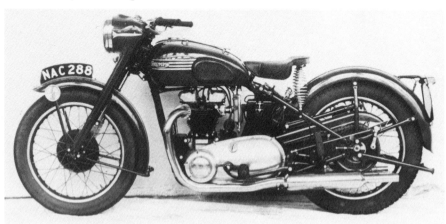

The Lucas alternator of the Speed Twin went on the end of the crankshaft with access via the cover in the chaincase.

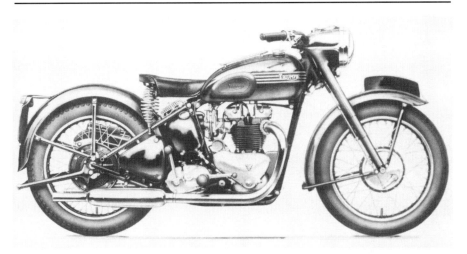

A 1954 Thunderbird, the first year it had coil ignition and an alternator plus the bigger timing-side main bearing.

removal. There was a lifting handle on each side of the rear mudguard and a rear stand that was pivoted from the frame.

A prop stand was listed as a general extra in the 1938 brochure, but not included for the Speed Twin in the parts list. It would, of course, have been easy enough to use that listed for the other models, as was to happen for 1939. A pillion seat and footrests were among the options, along with a rear carrier and a stop light, while a saddle was provided for the rider.

The oil tank went beneath the saddle on the right-hand side of the machine and was balanced by the six-volt battery and its carrier on the left-hand side. A triangular tool-box went to the rear of the oil tank between the chainstays. The petrol tank held 3¹/₂ gallons of fuel and was fitted with an instrument panel in its upper surface. This carried the oil pressure gauge, ammeter, light switch and an inspection lamp, but the speedometer went on the top of the front fork girder. Thus, it was easily driven from the front wheel, but suffered the full impact of all road shocks. In this aspect, the speedometer was joined by the 8 in., chrome-plated, headlamp shell, which was also fork-mounted.

The finish for the Speed Twin was the famous Triumph Amaranth red, which was applied to all the major painted parts. The petrol tank was chrome-plated first and then given top and side panels in the Amaranth red, these being lined in gold. The mudguards were gold-lined on each side of their central, raised rib, and the wheel rims were chrome-plated with gold-lined, Amaranth red centres.

The Speed Twin was very well received by both press and public, for its looks and style were so akin

Line of policemen with their Speed Twins at the Crystal Palace road racing circuit in 1959 where they practiced security for President Eisenhower's visit to Britain that year.

to a twin-port single and it performed well. During 1938, its potential was emphasised when a supercharged machine, ridden by Ivan Wicksteed and tuned by him and a friend, set the all-time Brooklands lap record for 500 cc machines at just over 118 mph.

Of more help to the firm was the trial by the London police of the Speed Twin for various duties in and around town. The machines proved very successful, and over the years many thousands were bought to serve the police well. In later years, the police had to turn to larger-capacity and faster models to keep pace with the traffic, but usually retained some Speed Twins for general use.

There were few changes for 1939, and most that came were of a general nature, such as the surround for the front number plate, which was to remain a Triumph trademark for many years. The primary chain-

The pivoted fork frame was introduced for the tourers in 1955 but this is the 1956 Speed Twin.

case was amended to provide a drip oil feed for the rear chain, a needle valve being fitted to control it. The major engine change was to increase the thickness of the block flange to strengthen it, and to use eight studs to hold it down. The original arrangement had proved marginal, but the new one was to continue from then on.

The Speed Twin was joined by the sports Tiger 100 for 1939, this having a raised compression ratio and the stylish silver and black Tiger finish. The two models were used in an attempt to win the Maudes Trophy early in 1939, this being awarded for the best demonstration by a firm of reliability, economy and easy servicing of its products. The two twins were selected from different dealers, run up and down the country, and then round Brooklands for six hours. Both machines came through with flying colours and the firm won the Trophy although, by

the time it was announced, the war had begun and it was of no account.

A Triumph range was announced for 1940, and the twins had several minor improvements. In the engine, the oil feed into the crankshaft was improved by adding a piston seal between the bush and the pressure valve. On the outside, the gearing was raised with a larger engine sprocket, and the frame was improved by a change of head angle and modified girder forks with a small check spring on each side. The petrol tank became the 4 gallon type used by the Tiger 100, and its instrument panel was made in steel instead of Bakelite. A slimline speedometer cable was introduced in a style to be used in the postwar years.

There was to have been a 350 cc twin for 1940, but this did not come about due to the war. The intended model was to have been the Tiger 85, and its announcement did reach a magazine front cover, but not its

A Thunderbird from the 1955-56 period having the fork bushes checked for wear.

editorial pages. A touring version was planned as the 3T, and the work done for these two models was used during the war to produce a 350 cc twin for the services. Plans to produce this in quantity had to be dropped, however, when the factory in Coventry was destroyed in 1940, after which the firm built single-cylinder machines elsewhere until the end of the war. By then, they had relocated to a new factory at Meriden, where all their post-war production was centred.

Postwar and 3T

Triumph were first off the mark with a post-war range, which was announced in March 1945. Initially, this comprised one single and four twins, but the single was never built, and only three of the twins appeared. The four twins had been listed as the touring 3T and 5T, and sports Tigers 85 and 100, but the smaller Tiger never did appear.

The Speed Twin picked up where it had left off, but with a number of basic changes. The most obvious of these was the use of telescopic front forks, based on a wartime design, with slim looks and hydraulic damping. External fork springs were used, but these were out of sight under long shrouds, which included lugs to carry the headlamp shell. This was reduced in size to 7 in.

Engine alterations mainly affected the electrics, as the mag-dyno was replaced by separate items. The magneto remained in the same place, but became either a Lucas K2F or a BTH type, flange-mounted to the back of the timing chest. With ei-

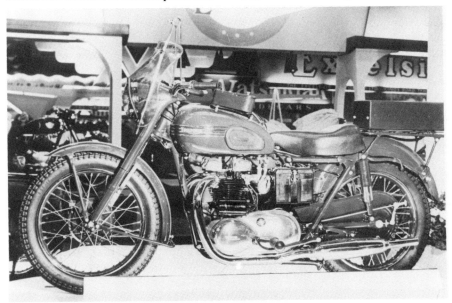

Earls Court in November 1955 with a police Speed Twin on show and fitted with both alternator and dynamo amongst its equipment.

ther make, the ignition was given auto-advance, the mechanism for this being incorporated within the driving gear. The mechanism itself varied, Lucas using bob-weights, and BTH a design with five rollers in a cage.

Current generation continued to be by a Lucas dynamo, but this was

removal of the dynamo.

Other engine changes were to the rocker box lubrication, the oil supply being taken from the return line. The external drain pipes were dropped and were replaced by internal drillings running down from the wells and through the block. The engine breather was changed

Neale Shilton showing the Lord Mayor of Coventry the police radio as fitted to one of the Triumph twins.

moved to the front of the crankcase, where it was clamped to the casting. It was driven from the exhaust camshaft gear, which meshed with a small pinion on the armature, and the timing chest was extended forward to enclose this. A hole in the rear of the chest accommodated the drive and was large enough for the pinion to pass through to allow easy

from the pre-war, one-way valve set in the crankcase wall to a timed disc valve driven by the left-hand end of the inlet camshaft. During the year, the connecting rod design was amended to separate bolts, rather than the integral studs of pre-war, and was to remain in the more usual style from then on. The carburettor became a Type 276 Amal, but the

The Thunderbird, as used by the TT marshals in 1956, here on trial with Chief Marshal, Peter Crebbin, on board.

size remained as before at 15/16 in.

The front wheel was reduced in size and its tyre changed to 3.25 x 19 in., while the oil tank was at first of 8 pints capacity, as for the Tiger 100, but later reverted to the 6 pints of pre-war days. The speedometer drive was moved to the rear wheel, and alternative instrument heads were shown in the parts list. One was the stock 120 mph type, but the other had additional scales on the dial to indicate the engine speed in the upper three gears. This special speedometer head was another Triumph feature that remained with the firm for many years.

The new tourer was listed as the 3T and never did acquire a model name. At first sight, it seemed to be a Speed Twin with a smaller capacity but, although it did have many parts in common with the larger models, it had a good few changes in its engine construction and its own frame. The dimensions of the 3T engine were 55 x 73.4 mm, which gave it a 349 cc capacity, and its flat-topped pistons resulted in a 7.0:1 compression ratio.

Inside, its crankshaft was unconventional, as it was built up to allow the use of one-piece connecting rods with white-metalled big-ends. The crankshaft comprised two halves and a central flywheel, so in this

Line of Thunderbird machines out in Fiji where they were used by the local police, just as with many other forces worldwide.

respect it was similar to the Speed Twin, but the way in which they were held together was totally different. Each half was forged as one part comprising mainshaft, crank cheek with bob-weight and crankpin. The crankpin of each forging was extended into the central flywheel, where it was clamped by a cross-bolt. During assembly, alignment was achieved by using a ground bar passing through holes in each half and the flywheel.

This crankshaft construction worked well enough in practice, and the assembly ran in a ball race on the drive side as usual, but in a flanged bush on the timing side. The remainder of the bottom half of the engine was essentially as for the Speed Twin, although a number of detail parts differed.

For the top half, it was the one-piece cylinder head that was most altered, as this was cast in iron with integral rocker boxes. It followed the form of the Speed Twin in general layout, so each rocker box housed two rockers on a single spindle. Each box had a single access cover retained by a knurled nut.

The cylinder head retained the splayed exhaust ports of the larger model but, unlike it, had the inlet manifold cast integrally. The manifold carried a Type 275 Amal carburettor of 7/8 in. bore with a separate float chamber. The head-to-block joint was sealed with two ring gaskets, each of which was fitted to a block spigot, and both head and block were secured by long through-studs and bolts. There were still eight fastenings, as for the larger engine, but the studs that were

Thunderbird, plus massive legshield, as for 1957 with the new tank badge common to the range.

The model Twenty-One as introduced early in 1957 with its bathtub rear enclosure and unit construc

listed as the 3TA and the forerunner of a line of models.

Ivor Davies of Triumph watches a 1957 Speed Twin with full-width front hub being placed on to its centre stand.

screwed into the crankcase top face extended up through the block to meet the bolts inserted from the top of the head.

The 3T had the standard gearbox and the stock clutch, with one less of each type of plate, but it had its own version of the primary chaincase. The frame was special to the model, although similar in type, but the forks, wheels and most other details were essentially the same. Differences amounted to the overall gearing and the fitment of a 3.25 x 19 in. tyre on the rear wheel, while the capacity of the petrol tank was only 3 gallons.

The finish of the Speed Twin remained as it was at first, but during 1946, the headlamp shell lost its chrome-plating and became painted in the same Amaranth red as the other details. The 3T copied the style of the Speed Twin, but was in black with the tank and mudguard lining in ivory, while its wheel rims were all-black.

The machines continued as they were for 1946, but for 1947, there were some detail changes and one major extra option. This option was the Triumph sprung hub, which had first been tried out in 1938, with a patent applied for in 1939 and the intention to use it on the 1941 range.

The sprung hub was an ingenious idea to offer very limited rear wheel movement (some 2 in. maximum) while retaining the standard rigid frame, and to do this at minimal cost and added weight. In this, Edward Turner succeeded, but the design was undamped, other than by friction, and it relied on some accurately machined parts to guide the wheel up and down. The springs inside were under considerable compression, so that dismantling was dangerous if not carried out with the correct tools. With the sprung hub fitted, the handling of the machine left something to be desired.

Due to the size of the sprung hub, its rear brake was increased in diameter to 8 in., and its use prevented the speedometer drive being taken from the rear wheel. To overcome this, a drive was contrived from the gearbox output sprocket, and the drive box was bolted to the gearbox shell.

Other changes to the machines were minimal, as production was paramount at that time. The float chamber of the 5T was moved to the left-hand side of the mixing chamber during this period, and the headlamp glass became domed. The finish for both models was unaltered, except when material shortages affected the amount of chrome-plating. A prop stand was offered as an option and clamped to the lower frame tube on the left-hand side. A lug on the clamp located the stand to prevent it turning on the tube, and in this respect the details varied between the two models.

Changes for 1948 were minimal, and for the Speed Twin concerned the rear mudguard area. The guard itself was amended so that the joint between the two parts moved round to the upper chainstays. Thus, removal of the tail section took away most of the guard to leave the wheel well exposed and easy to attend to. In addition, the lifting handles were deleted and the top of the rear number plate formed to do this duty.

Police Speed Twins out in Hong Kong and parading before H.E. the Governor, Sir Alexander Grantham, shortly before his retirement.

None of the mudguard changes applied to the 3T but, during the year, this model had its engine modified a little. The alteration was to the head and block fixing, which became similar to that of the 5T, so there were short studs to hold the block, bolts to secure the head to the block, and a single head gasket. The finish of the 3T was altered only in that the wheel rims were chrome-plated and finished with black centres lined in ivory.

Rather more was changed for 1949, as that was the year the famous Triumph nacelle first appeared to tidy up the area around the top of the forks, the handlebar clamps and the headlamp mounting. The top of the nacelle carried the speedometer, plus the light switch and ammeter from the tank panel, and the ignition cut-out from the handlebars. The light unit went into the front of the nacelle with the horn just below it, but out of sight behind slots cut in the lower nacelle pressings.

With the appearance of the nacelle, the tank panel went and with it the oil pressure gauge. The gauge was replaced by a tell-tale, incorporated in the pressure release valve set in the timing cover, and the tank top was given the option of a parcel grid. The grid proved very popular and was soon fitted as standard, although it could be removed, and Triumph thoughtfully provided four plugs to fill the tapped holes in the tank top.

The handlebars were altered to suit the nacelle and tidied up by having the horn button screwed into them and the dip-switch mounted on the back of the front brake lever block. The petrol tank capacities remained at 3 gallons for the 3T and 4 gallons for the Speed Twin, while both had the speedometer driven from the gearbox sprocket for all production machines, whether they had a sprung hub or not. All had an air filter fitted as standard, and the Speed Twin engine went over to a roller race for the timing-side main bearing. The finishes were unaltered, and this brought the tourers to the end of a decade and the point where a new model was to join them.

Thunderbird and sprung frame

Once the worst of the war was past, a demand began for more performance for sporting events, especially from the USA. At first, this was met by the Tiger 100 but, although this responded well to tuning, the American adage that there was no substitute for increased capacity was often voiced.

Under this pressure, Triumph, which meant Edward Turner who spent considerable periods of time in the USA, agreed to increase the capacity of the twin, and the result was spectacularly launched late in 1949 as the Thunderbird, or 6T. It was closely modelled on the Speed Twin in both engine and chassis, with all machines in the range benefiting and adopting the changes it brought with it.

The Thunderbird engine was

First of the small unit construction twins was this 1957 Twenty-One model with bathtub rear enclosure.

based on 71 x 82 mm dimensions, which gave it a 649 cc capacity. The compression ratio was 7.0:1 and the carburettor a Type 276 of 1 in. bore. It copied the Speed Twin in many respects and shared a good number of common parts, although the major components were special to it. Due to the increase in bore size, it was no longer feasible to drill the oil drain holes in the block, so the external drain pipes were used for the 6T and

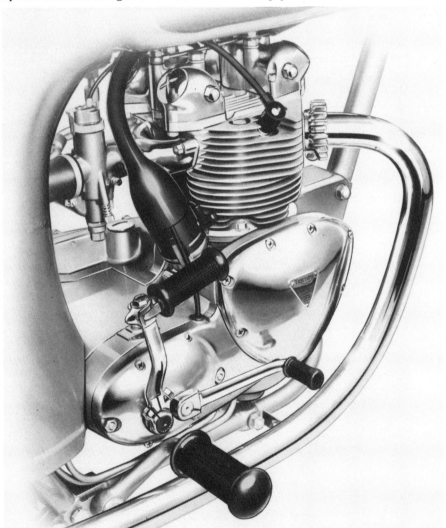

The 349 cc engine unit of the Twenty-One showing the distributor tucked in behind the cylinder block.

Drive side of the Twenty-One which was listed as the model 3TA soon after its launch.

returned for the 5T.

All engines were fitted with a higher-output dynamo, and the extra power of the Thunderbird called for a stronger gearbox.To meet this need, the gearbox was redesigned and, at the same time, the speedometer drive was taken from the end of the layshaft. The new gearbox was fitted to all models, which also had the lug for the prop stand incorporated into the frame and a dualseat option. The sprung hub continued to be listed, but in a MkII form with better bearings.

The petrol tank finish was changed completely, and in place of the chrome-plating and lined panels were painted tanks with chrome-plated styling fittings on each side. These fittings comprised four hori-

zontal bands, with the Triumph name badge attached to them, and greatly reduced the plating, which had been giving trouble on exported models. The tank capacity for the 6T was 4 gallons, the same as the 5T, while the 3T kept to its 3 gallons. The finish of the smaller models continued as it was, but with the all-over paint scheme for the petrol tank, and the 6T was similar, but in blue-grey with gold lining.

The Thunderbird launch took place at the Montlhéry track near Paris, where three machines were set the task of covering 500 miles each at over 90 mph. In this, they were successful, as the running average for all three was over 92 mph while, when stops were included, two were over 90 mph and one at 86

The dualseat of the Twenty-One hinged up to expose the moulded toolpad, the electrics and the oil filler cap.

mph, this last having had a couple of unscheduled stops. All three also did a flying lap at over the one hundred mark and were then ridden back home.

There were only minor alterations for 1951; a bayonet petrol cap, tapered rear light of larger area, and cast-iron front brake drum. The carburettor of the Thunderbird was increased in size to 1-1/16 in. while the 3T petrol tank capacity went up by a 1/2 gallon. The finishes remained in the same style, and their only change was to a lighter polychro-

matic blue for the 6T. At the end of the year, the 3T was dropped, as the firm concentrated on the larger models.

The frame for the 5T and 6T was altered for 1952 to include a seat tube lug with an eye in it. This allowed the air filter, also new, to fit behind the tube and be connected to the carburettor with a straight piece of hose. The Thunderbird had a complete change in this area, for it went over to an SU instrument, type MC2, which meant a new inlet manifold to suit. It was a type with which

The original pre-unit Speed Twin came to an end in 1958 when this one received its final pre-sale check over.

motorcyclists were not very familiar, although it was common enough on cars.

One further change for 1952 concerned the nacelle and headlamp. The light unit became the sealed-beam type without a pilot bulb, so a small rectangular lamp was fitted over the nacelle grille area. The type was known as the underslung lamp, and was both ineffective and unloved. The machines' finish remained as it was in 1951 except where plating restrictions, brought about by material shortages, affected it. Due to this, the handlebars and wheel rims were painted, and some details finished in cadmium- rather than chrome-plating for part of the production run.

For this 1958 Thunderbird it was the last year of the SU carburettor and the first for the Slickshift.

Early Twenty-One fitted with matched panniers and making a nice display.

Little was altered on the Thunderbird for 1953, but the Speed Twin went over to coil ignition and was fitted with a Lucas RM12 alternator on the left-hand end of the crankshaft. Its presence meant that there was no longer any room for the engine shock absorber, so this function was moved into the clutch centre for all models. Where there was no alternator rotor, a distance tube was fitted to fill up the space, and an access cover appeared in the primary chaincase.

The ignition system of the Speed Twin was based on a single coil mounted above a points-and-distributor assembly, which replaced the magneto and included the auto-advance mechanism. It was driven by a simple spur gear, which was held to the shaft by a cross-pin re-

The 1959 Thunderbird which fitted a Monobloc carburettor for the first time that year.

tained by a circlip. An ignition switch joined the lighting switch and the square Westinghouse rectifier went under the seat.

Both the Speed Twin and Thunderbird had new camshafts with ramp cams, which made for quieter running, and the 5T crankcase was amended to remove the dynamo mounting and provide for the alternator. A rectangular rear lamp with red plastic lens went on both models, which continued with no change to their finish.

Edward Turner on a 5TA Speed Twin leads a party of industry and press riders out on a run on a variety of Triumph models.

Police riders with their unit construction Twenty-One machines each fitted with legshields and a special dualseat.

For 1954, the Thunderbird was fitted with coil ignition and a Lucas RM14 alternator, which went on the Speed Twin as well. Both models changed to a round rectifier and had the ignition switch built into the lighting one, with a central ignition key, and both were fitted with a new

Police Thunderbird for 1960 when it still had magneto ignition but gained the new bathtub.

The 1961 Thunderbird had lighter lines thanks to the two-tone paint finish but retained much from the past with the new duplex frame.

style of silencer of barrel shape. The 6T joined the 5T in having an access cover in the primary chaincase, using the same detail parts.

For the 6T alone, there was a new crankshaft with an increase in crankpin and timing-side mainshaft diameters. This brought in a new connecting rod, while the timing side main bearing was increased in size and changed to a ball race. Its larger outside diameter required an amendment to the crankcase casting, which produced a bulge under the timing chest, making it easy to pick out this type. The colours continued as they were.

There were more radical changes for 1955, when the Speed Twin was fitted with the larger crankpins and timing-side main bearing. Its carburettor was changed to a Type 376 Monobloc of 15/16 in. bore, but the 6T continued with its SU. The real change affected both models and was to a new frame with pivoted-fork rear suspension. This had been introduced for the Tiger models in the previous year and, essentially, was common to the range. The design involved a press-fit for the rear fork pivot, which made it awkward to replace the pivot bushes, but it was a real step forward from the sprung hub.

The cycle parts were amended to suit, with a centre stand in place of the rear one, a new oil tank and a matching combined toolbox and battery carrier. A dualseat was fitted as standard, the access cover in the chaincase disappeared, and the rear brake backplate was restrained by a torque stay. A quickly-detachable rear wheel was listed as an option, the rear lamp lens incorporated a rear reflector, and the rectifier was reduced in size. However, the finishes remained in the Amaranth red and Polychromatic blue, as before.

After this major change, there were few alterations for 1956. In the engine, the connecting rods were fitted with replaceable shells to make servicing easier for the owner, and the Speed Twin began to use the 6T crankcase, which became the common part. The clutch went over to bonded-on inserts. On the outside, the underslung lamp was replaced by a plated horn grille, as the pilot bulb was located in the headlamp reflector. The petrol tank parcel grid had its bars altered to suit the style of a central tank seam strip that had been added. The finish finally changed, but only for the Thunderbird, the colour of which became Crystal grey.

The changes for the 5T and 6T for 1957 were again minor, the most obvious being a new full-width front wheel hub. This was of composite construction, a cast-iron drum being riveted to a brazed assembly of steel pressings that supported it and held the wheel bearings. The brake continued to be a 7 in., single-leading-shoe type, and the other end of the hub was fitted with a cover plate. To go with the new front hub, there were forks with separate end caps to clamp on to the wheel spindle.

Bristol Triumph Owner's Club after a 24 hour run in the early 1960s; 5TA on left and 3TA centre.

A 1961 3TA out on the road where its well valanced front mudguard, bathtub and additional screen helped to keep the weather at bay.

The machines' looks were altered by a change to the tank badges, the new ones having a grille between upper and lower bars, plus the Triumph name. Associated with the new badges were styling strips that ran fore and aft of them to create a line around which a two-tone paint finish could be offered but this was not made available immediately for either model. The finish of the 6T changed, the frame, forks, oil tank, battery box and details altering to black and the petrol tank and mudguards to bronze-gold. The Speed Twin remained in its traditional Amaranth red and the wheel rims of both models were simply chrome-plated.

Early in 1957, the two-model touring range was increased in size, and the additional new machine set a new trend with its own unique style and line.

Unit construction

The new model for 1957 was called the Twenty-One, to celebrate the 21st birthday of the reconstructed 1936 company, and because its capacity was given as 21 cubic inches in the USA. It broke new ground for a production Triumph twin, as it had unit construction of engine and gearbox, plus a considerable degree of rear enclosure and good weather protection. It still retained its Triumph line and looks so, once again, the firm managed to step forward without seeming to move too far from its established customers.

In time, the machine became listed as the 3TA, and its engine layout was really little altered from that of the Speed Twin. Its dimensions were 58.25 x 65.5 mm, and the capacity 349 cc. The compression ratio was 7.5:1 and the carburettor a Type 375 of 13/16 in. bore. Ignition was by coil, the points assembly,

The final year for the pre-unit Thunderbird was 1962 when it lost the Slickshift and was fitted with a siamesed exhaust system.

For 1963 the Thunderbird changed to unit construction and a skirt in place of the bathtub. There was also a new frame.

auto-advance and distributor being in a single housing to the rear of the right-hand cylinder, where it was driven by skew gears from the inlet camshaft.

In most other aspects, the 3TA followed the older twins, but it did have a light-alloy cylinder head with the cast-iron block painted silver to match. The top-half fixings, separate inlet manifold, twin rocker boxes and pushrod tubes were all in the Speed Twin mould, but there were no external oil drain pipes, and the exhaust port stubs were screwed into the cylinder head.

The bottom half differed in that the crankshaft was forged in one piece, the flywheel being pressed on to the centre cheek and secured by three radial bolts. The connecting rods were steel, but kept to shell big-end bearings, and the crankshaft turned in a ball race on the drive side and a bush on the timing side.

The timing chest and oil pump were as for the larger engines, although the detail parts were generally new, but the camshafts ran directly in the crankcase at their right-hand ends and were retained by plates bolted to the case. The inlet camshaft continued to drive the crankcase breather and oil pump, as well as carrying out its new task of driving the distributor. A Lucas type RM13/15 alternator went at the left-hand end of the crankshaft, with a duplex primary chain inboard of its rotor.

The crankcase halves were very similar in layout to those of the pre-unit engines, but they were extended to encompass the transmission. The joint line still ran through the centre of the engine, but then stepped over to the left so that the entire gearbox shell formed part of the right-hand casting. Inner and outer covers completed the assembly, the latter being shaped to blend to the rear of

The police version of the 1963 Thunderbird had 12 volt electrics, a year ahead of the standard models, as well as its other special equipment.

the timing chest. The timing cover was of triangular shape and smaller, as it no longer had to enclose the magneto drive, but it retained the Triumph line. The left-hand crankcase casting included the inner half of the primary chaincase, in which there was a large hole to give access to the gearbox sprocket. A circular plate filled this hole and carried an oil seal, which ran on the gearbox mainshaft.

The clutch was much as for the other models, with a shock absorber in its centre, but the chainwheel was to suit the duplex primary chain. The four-speed gearbox differed in having the layshaft behind the mainshaft, and both were mounted, with the gears, in the inner gearbox cover. The positive-stop and quick-thread clutch mechanisms went between the covers, while the kickstart pedal worked a pawl and ratchet inside the layshaft first gear.

The complete engine unit went into a frame with pivoted-fork rear suspension and bolted-on subframe. It had telescopic front forks in Triumph style, with nacelle and steering damper, and both wheels had 3.25 x 17 in. tyres and 7 in. brakes, the front in a full-width hub. The speedometer drive was taken from the rear wheel.

It was the rear enclosure that set the 3TA apart from its contemporaries, for twin panels were bolted together to run from the dualseat nose to the rear number plate. The bulk of the rear wheel could still be seen, so the fitting did not look at all heavy, but it gave the machine a good line. The enclosure had the appearance of an inverted hip bath so, invariably, it became known as the 'bathtub', a name that was, and still does, apply to all models so fitted.

The bathtub was very sleek, as its

joint flange was turned inward, and under it went a simple mudguard. The dualseat was hinged to the frame, but it sat on the bathtub top with a rubber toolpad under it, this being moulded to carry all the items of the tool kit. At the front of the machine, the rear enclosure was matched by a very deeply valanced mudguard, and the two were set off by a $3\frac{1}{2}$ gallon petrol tank fitted with the new grille badges. The finish of the show model was in silver-grey, but for production, the tank, forks,

too popular with owners. Other changes for these machines were to add an oil seal in the engine outboard of the drive-side main bearing, and a steering head lock to the headstock.

During the year, the 3TA carburettor size went down to 25/32 in., while the finish of all three models remained unchanged. At the end of the year, the pre-unit Speed Twin finally came to the end of its 20 year life, having set the standard for the industry, but now being outmoded

The 1964 3TA with points in the timing cover and skirt in place of the bathtub.

front mudguard and rear enclosure were in Shell blue and the other painted parts in black.

The three models ran on for 1958, with a gearbox alteration being the main change for the pre-unit machines. This introduced the 'Slickshift', in which the action of changing gear also lifted the clutch thanks to a linked mechanism. The clutch lever remained for moving off and stopping, but the feature was none

and due to be replaced.

The new Speed Twin was the logical development of the 3TA and appeared for 1959 as the 5TA. It shared a great number of common parts with the smaller twin in both engine unit and chassis, while the 490 cc capacity came from 69 x 65.5 mm dimensions. The compression ratio was 7.0:1 and the carburettor a Type 375 of 7/8 in. bore. The head, block, pistons and valves differed between the engines, the larger twin having light-alloy connecting rods and different pushrods. Just about everything else in the engine was common, as was the clutch and gearbox. It was the same with the cycle parts, except for the rear tyre, which was increased in section to 3.50 x 17 in. The finish was in the traditional Amaranth red for all painted parts, while the 3TA remained in its Shell blue and black.

The Thunderbird had its colour changed for 1959, when the petrol tank, mudguards and front forks were in Charcoal grey. The SU carburettor was finally dropped in favour of an Amal Type 376 Monobloc of 1-1/16 in. bore, while the alternator type became a Lucas RM15. Internally, its crankshaft construction was changed to the one-piece shaft and three-bolt flywheel, as used by the unit twins.

If 1959 had been the year of change

for the Speed Twin, 1960 was that of the Thunderbird. New for the big twin was the frame, which had duplex downtubes, and the model also adopted the bathtub rear enclosure and well valanced front mudguard from the smaller models. The frame was introduced with a single top tube, but there were a few breakages, so it was quickly modified to incorporate a brace under the top tube.

There was an amendment to the bathtub mating flange, which was turned out instead of in, for all three models. The 6T was fitted with the hinged dualseat, but not the moulded toolpad, its toolkit going into a tool roll. Finally, the 6T wheel size was reduced to 18 in., and the tyres to 3.25 x 18 in. front, and 3.50 x 18 in. rear.

For the small twins, there was just the change to the bathtub flange, the option of a quickly-detachable

The Thunderbird of 1964 with revised front forks and 12 volt electrics as standard.

For 1965 the Thunderbird was fitted with a sports front mudguard but little else was altered.

rear wheel, and a primary chain tensioner for the Speed Twin alone. The 1960 finishes were as before for the 3TA, but in Ruby red in place of the Amaranth for the 5TA, a brighter shade after over two decades without alteration. For the 6T, the Charcoal grey continued and was extended to included the bathtub.

There were floating brake shoes for all for 1961, to improve the brake efficiency, while the 3TA gained the primary chain tensioner. Both 3TA and 5TA lost the toolpad, while the tank styling bands became optional. The Thunderbird had rather more changes, for it was fitted with a light-alloy cylinder head and had its compression ratio raised to 7.5:1. The external oil-pipe drains from the valve wells were no longer fitted, and in the gearbox there were needle races for the layshaft. On the outside went a folding kickstart pedal, and the front brake went up to an 8 in. size, but remained in the full-width format of the past.

The finishes for the smaller twins were unchanged, but the 6T went to a two-tone effect. In this, the lower petrol tank, front mudguard and bathtub were in silver, as were the fork legs, while the rest of the painted parts were in black, including the nacelle and upper fork legs. The lining was in gold and the wheel rims were chrome-plated.

The Thunderbird lost the Slick-shift gearchange for 1962 when, together with the two smaller twins, it was fitted with a Lucas RM19 alternator. All three also went over to a siamesed exhaust system, with the single silencer mounted low down on the right. The finish of all three models remained as for the previous year.

There were major changes for the Thunderbird for 1963, for it went over to unit construction of the engine and gearbox, a new frame and reduced rear enclosure. The

The smaller twins also fitted the sports front mudguard for 1965, this being the 3TA.

revised engine adopted the layout of the smaller twins for its crankcase, but kept to its original layout for most of the internals and the top half.

Many detail parts were amended to suit the changes, so relatively few remained common. Thus, the crankshaft lost its timing-cover bush support, which was replaced by an oil seal. This reduced wear and improved the oil feed to the big-ends. A ninth bolt was added to the cylinder head fixings, while each rocker box cap lost its hexagon, but gained a double coin slot and a locking tab.

The most obvious engine change was the move of the ignition points to the timing cover. Their cam and its auto-advance mechanism were driven from the exhaust camshaft, and they fired twin coils. The ignition switch became separated from the lighting one, but both remained in the nacelle with the ammeter and speedometer.

The primary drive was altered to a duplex chain with a tensioner under the lower run. The clutch was changed to a type with only three springs to clamp the plates, but was otherwise very much as before. Due to the reduction in number of springs there had to be a corresponding change to three sets of shock absorber rubbers. The gearbox remained the four-speed type of old.

Bolted construction continued for the frame, but the new front section had a single downtube and much improved support for the rear fork. This now sat between the ears of a lug, brazed into the seat tube, so that its removal became much easier.

The remainder of the Thunderbird was much as before, but the exhaust system went back to twin pipes and a silencer on each side of the machine. The petrol tank continued to carry 4 gallons of fuel. Brakes were as before, with an 8 in. one at the front in a full-width hub, and a 7 in. one at the rear in an offset hub, a quickly-detachable hub still being an option. The bathtub did change to become a rear skirt, the line of which ran up the subframe tube to a valance that ran round the

went on to the smaller twins, which were fitted with the timing-chest points, rocker box cap locks, three-spring clutch, twin switches for lights and ignition, and a folding kickstart pedal. They also went over to a 3 gallon petrol tank and to the skirt for their rear enclosure, so the bathtub was no more. All three models had revised front forks, and the 6T went over to 12 volt electrics.

The finish of the 3TA became sil-

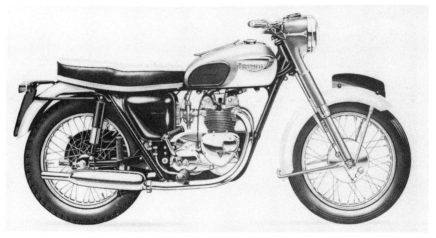

Final year for the smaller unit twins in touring guise was 1966 when they lost the skirt as shown by this 3TA.

rear of the dualseat.

Little happened to the smaller twins, other than a reversion to twin exhaust pipes and silencers. Their colours were amended so that the 3TA was given an option of the Shell blue parts being in silver-bronze, and the 5TA colour became known as Cherry red. For the 6T, the colour continued in silver, but extended to the rear mudguard and skirt in place of the bathtub.

For 1964, most of the changes

ver-bronze as standard for the tank, forks, mudguards and skirt, with the rest of the painted items in black. The 5TA and 6T went to a common two-tone finish, the lower tank, mudguards and skirt being in silver. The tank top and other painted details were in black, but the lower fork legs were in silver.

Engine changes were minimal for 1965, when all three models lost their oil tell-tale, the 3TA had its port size reduced, and the 6T had a top-dead-centre slot machined into its flywheel. All models lost the surround to the front number plate, which had been a Triumph feature for so long, and the well valanced front mudguard was replaced by a sports style. The smaller twins had their frame revised to include a tubular strut between the headstock and the seat nose. This addition brought in a change to the petrol tank mountings, and the tank capac-

oil tank for the rear chain.

The smaller twins had the frame strut welded in place and went over to 12 volt electrics. Their wheels became common for both models and increased in size, so the front one was shod with a 3.25 x 18 in. tyre, and the rear one with a 3.50 x 18 in. cover. The Thunderbird engine was fitted with a roller bearing main on the drive-side and given an oil feed to the exhaust tappets. The

The final Thunderbird also lost its skirt for 1966, after which its role was taken over by the TR6; 1966 was also the last year of the famous Triumph nacelle.

ity returned to 3½ gallons. All the models kept their existing finish, but that of the 3TA was renamed silver-beige.

The final year of the touring Triumphs was 1966, for which all three had a number of changes. All lost the rear skirt, for trends were away from enclosure, and were fitted with a small panel on the left-hand side, to match the oil tank, new badges, of 'eyebrow' style, for the petrol tank and an oil feed from the

6T frame had fairing brackets added to the headstock, and its front hub gained a spoke flange so that the brake could be fitted with wider shoes. The 6T speedometer drive was taken from the rear wheel as on the smaller twins, so the gearbox drive was deleted.

All three models had a two-tone finish, with the 3TA in Pacific blue for its petrol tank top and front forks. The lower tank and mudguards were in Alaskan white, while the rest of the painted parts were black. The 5TA kept its 1965 silver and black finish, minus the rear skirt and with the left-hand side cover in black. The 6T was similar to the 5TA, with silver for the lower tank and mudguards, and black elsewhere, this now including the fork legs. For the USA, the 6T tank finish was in Grenadier red and Alaskan white, while all three models were fitted with white handlebar grips that year.

That brought the tourers to their end, for all three models were dropped at the end of 1966 as the firm concentrated on its sports machines. At the time, it seemed the right commercial decision, but many riders mourned the passing of the models that had done such sterling service for so many years.

Triumph Touring Twins Specifications

All models have twin cylinders, overhead valves and a four-speed gearbox

Model	3T	5T	5T	5T	3TA	3TA
years	1945-51	1938-40	1945-54	1955-58	1957-65	1966
bore mm	55	63	63	63	58.25	58.25
stroke mm	73.4	80	80	80	65.5	65.5
capacity cc	349	499	499	499	349	349
comp. ratio	7.0	7.0	7.0	7.0	7.5	7.5
carb type	275	76	276	376	375	375
carb size	7/8	15/16	15/16	15/16	13/16[1]	25/32
ignition by	mag	mag	mag[2]	coil	coil	coil
generator	dyn	dyn	dyn[3]	alt	alt	alt
voltage	6	6	6	6	6	12
top gear	5.78	5.00	5.00	5.00	5.33[4]	6.04
petrol - gall	3[5]	3.5	3.5	3.5	3.5[6]	3.5
frame	rigid[7]	rigid	rigid[7]	s/a	s/a	s/a
front tyre	3.25x19	3.00x20	3.25x19	3.25x19	3.25x17	3.25x18
rear tyre	3.25x19	3.50x19	3.50x19	3.50x19	3.25x17	3.50x18
front brake dia	7	7	7	7	7	7
rear brake dia	7[8]	7	7[8]	7	7	7
wheelbase in.	52.2[9]	54	54[10]	55.7	52.7	52.7

[1] - 1958-25/32
[2] - 1953-coil
[3] - 1953-alt
[4] - 1964-5.40, 1965-5.70
[5] - 1951-3.5
[6] - 1964 only-3
[7] - 1947-sprung hub option
[8] - 1947-8 with sprung hub
[9] - 1950-53.2
[10] - 1950-55

Triumph Touring Twins Specifications

Model	6T	6T	6T	6T	5TA	5TA
years	1950-54	1955-60	1961-62	1963-66	1959-65	1966
bore mm	71	71	71	71	69	69
stroke mm	82	82	82	82	65.5	65.5
capacity cc	649	649	649	649	490	490
comp. ratio	7.0	7.0	7.5	7.5	7.0	7.0
carb type	276[1]	SU[2]	376	376	375	375
carb size	1[3]	1-1/4[4]	1-1/16	1-1/16	7/8	7/8
ignition by	mag[5]	coil	coil	coil	coil	coil
generator	dyn[6]	alt	alt	alt	alt	alt
voltage	6	6	6	6[7]	6	12
top gear	4.58	4.58[8]	4.67	4.60	4.80[9]	5.40
petrol - gall	4	4	4	4	3.5[10]	3.5
frame	rigid[11]	s/a	s/a	s/a	s/a	s/a
front tyre	3.25x19	3.25x19[12]	3.25x18	3.25x18	3.25x17	3.25x18
rear tyre	3.50x19	3.50x19[13]	3.50x18	3.50x18	3.50x17	3.50x18
front brake dia	7	7	8	8	7	7
rear brake dia	7[14]	7	7	7	7	7
wheelbase in.	55	55.7	55.7	55	52.7	52.7

[1] - 1952-SU
[2] - 1959-376
[3] - 1951- 1-1/16
[4] - 1959- 1-1/16
[5] - 1954-coil
[6] - 1954-alt
[7] - 1964-12
[8] - 1960-4.47
[9] - 1961-5.05, 1964-5.13
[10] - 1964 only-3
[11] - sprung hub option
[12] - 1960-3.25x18
[13] - 1960-3.50x18
[14] - 8 with sprung hub